PONDMASTER

# An essential guide to choosing your
# POND FISH
## AND AQUATIC PLANTS

GRAHAM QUICK

D1298271

BARRON'S

## Author

**Graham Quick** has been keenly interested in ponds and fish from an early age. Even before leaving school, he spent his spare time working on a commercial fish farm. After leaving school he created an extremely successful aquatic center at his family-run garden center. He is now in the process of setting up an ornamental fish farm.

## Credits

Created and designed: Ideas into Print,
New Ash Green, Kent DA3 8JD, England.

First edition for the United States and Canada published in 2000 by Barron's Educational Series, Inc.
First published in 1999 by Interpet Publishing
Original edition © 1999 by Interpet Publishing

*All inquiries should be addressed to:*
Barron's Educational Series, Inc.
250 Wireless Boulevard
Hauppauge, NY 11788
http://www.barronseduc.com

International Standard Book No. 0-7641-5271-8
Library of Congress Catalog Card No. 99-69003

Printed in Hong Kong
9 8 7 6 5 4 3 2 1

*Below:* The elegant flower spikes of Primula vialii *create a colorful parade at the edges of this stream. This and other bog garden primulas thrive in moist soil and begin their flowering display in late spring.*

# Contents

*Shubunkins, a highly decorative goldfish variety, make attractive pond fish.*

# Choosing pond fish

For many people, fish are the best part of pondkeeping. They can add the finishing touches to any pond, but if you make the wrong choices, the species you introduce can either overpower the pond or be so inconspicuous that they make no impact at all. Choosing the right fish is vital, as much for the fishes' health as for the viewing pleasure they provide. There are a few things to bear in mind before introducing any fish to the pond. The first is stocking levels: allow 2 cm (0.8 in) of fish per 50 liters (11 gallons) of water. Always understock rather than overstock the pond. This allows the fish to breed up to the natural level; it reduces the risk of disease outbreaks; and improves the growth rates of the fish. A few fish need special foods and certain conditions to be seen at their best. Make sure you can meet these needs before you proceed, and never buy on impulse; many fish are totally unsuitable for ponds. If a fish grows too big for your pond, never release it into the wild; it could spread disease or displace native fish. Be sure to check local regulations on keeping certain fish species.

Start by introducing a few hardy fish and wait for a month or two before stocking up with expensive ones. This gives you a chance to see if there are any problems. Buy fish from just one or two retailers, as this will reduce the chance of introducing a disease that the fish have not encountered before. Quarantine new fish for four weeks before placing them in a pond already containing fish. The guidance given here will help you create a healthy environment for your fish. Bear in mind that the equipment and products featured will vary in different parts of the world.

When you first stock your pond, the most important objective is to buy healthy fish. Try to obtain locally bred fish, as imported ones will have had to endure the stress of transport and will not be accustomed to your local conditions. Very small fish measuring 5 cm (2 in) or less are normally poor travelers, so avoid them. Look instead for fish measuring 6-8 cm (2.4-3.2 in). Make sure that the first few fish you buy are hardy ones. Look out for damaged fins and growths on the body. A sick fish will stay on its own, away from other fish. Avoid buying either the one on its own or the fish housed with it. Fish should be active and swimming around. (Tench are the exception, as they huddle together in tanks.)

Ask the store assistant to bag up the fish and check them again before you pay. If you are not happy, do not buy them. Take the fish straight home; keeping them in a car for a couple of hours while you do your shopping is not the way to ensure healthy specimens. As soon as you arrive home, float the fish on the pond for 15 to 20 minutes to allow the temperature to equalize between the bag and the pond. Open the bag and let in some water from the pond. This allows the fish to adjust to your pond water conditions. Leave it for another 15 to 20 minutes and then release the fish by tipping the bag slowly into the water so that the fish can swim out.

It may be a few days before newly introduced fish come to the surface to feed. Allow them to settle for a few weeks before adding more fish. This will give you time to see if there are any problems. Remember to test the water to ensure that the filter can cope with the waste produced by the new fish.

*1 Take your fish home as soon as you can. The dealer will put them in a plastic bag with air (or oxygen for long journeys) taking up most of the space in the bag. Keep the bag cool and in a box or covered, as the fish are less stressed in the dark.*

*2 Leave the bag floating on the surface of the water. Avoid full sun, because this will heat the bag and stress the fish inside. After about 20 minutes, mix some pond water with the water in the bag, retie it and float it for 15 minutes more.*

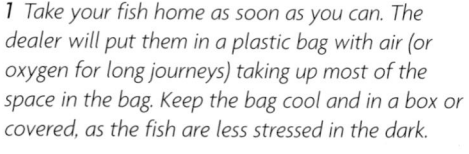

*3 After letting the fish get used to the water, slowly tip the fish and water out into the pond, making sure that all the fish are out of the bag before putting it away. (Keep the bag; you may need it if you have to transport a fish one day.) The fish will hide for a day or two, so there is no need to feed them. After this, feed them once a day to start with, then twice a day.*

# Feeding your fish

In the wild, fish feed all the time, so their digestive systems are designed to cope with small but frequent meals. So that the fish can derive maximum benefit from their food, follow a similar feeding pattern; the best plan is to feed them four or five times a day. If this is difficult, invest in an automatic fish feeder (which is also useful if you go on vacation). Always buy good-quality food and avoid loose products, as these rapidly deteriorate in air. Use two kinds of food at the same time to give the fish some variation and a more balanced diet. To vary the diet even more you can feed brown bread, bloodworms (frozen are safer as fresh can carry disease), lettuce, spinach, and even peas.

*You can buy frozen foods from most pet stores, the most common being bloodworm (shown here), daphnia and black mosquito larvae. Keep in the freezer until needed, break off only what you need and defrost it in cold water before feeding it to the fish. Do not feed frozen foods directly to the fish.*

*Pond flake is the ideal starter for small fish up to about 5 cm (2 in). It is easy to digest and as the flakes spread across the pond, all the fish get a chance to eat some.*

*Brown bread offers an additional food source for fish. It is high in minerals and is easy to digest, especially in the winter months.*

*Once the fish have grown to about 5 cm (2 in) or more they should be able to eat sticks or pellets, which will provide more food per mouthful than flake. The fish still need four or five feedings per day.*

## Feeding table

**Spring** As the water temperature rises over 10°C (50°F), start to feed small amounts once or twice a day.

**Summer** Now the fish should be eating four or five times a day to spur growth and prepare them to spawn. If you go away during the summer, ask a friend or neighbor to feed your fish regularly according to your directions.

**Autumn** The fish will need to eat a lot of food to put on weight to survive through the hibernation over winter.

**Winter** If the temperature drops below 10°C (50°F), the fish will not be very interested in food, but they may take food on milder days. If they are looking for food, brown bread is good—but only once per day.

Why filter your pond ? The main reason is to improve the water quality so you can stock more fish and enable them to grow and prosper. Good water quality also means that the pond will be clear enough to make the fish visible all year round. With suitable equipment you can create a pond of a size and in a position that in nature would not work, such as a very small pond or one in full sun with no plants. Using pumps and filters to sustain a pond brings with it the responsibility of maintenance, perhaps on a daily basis. It is vital to check that all the equipment is working and that there are no leaks. If you go away, it is worth asking a friend to check every day that nothing goes wrong. Show them what to do if there is a problem. Whenever you work on the pond, be sure to turn off the electricity supply.

## **A typical pond filter system**

*The UV sterilizer greatly reduces the population of ⌐d some parasites.*

*The biofilter strains out suspended matter and breaks down toxic waste by bacterial action.*

*The water returns down a waterfall.*

*Pump draws in pond water and pushes it around the system.*

## **Maintaining the pump**

The pump is the easiest thing to check, because when it is blocked the water stops flowing. It is also the most neglected piece of pond equipment. Once a week, if not more often, remove the strainer and sponge and clean them in fresh water. At the same time, inspect the impeller for dirt or any foreign objects. If it is dirty, take it apart and clean it. In hard water areas, the pump may need descaling, perhaps as often as every six months. Follow the manufacturer's instructions.

*Water is drawn into the pump through this plastic strainer, which excludes any particles that could block the pump impeller.*

*Main pump body and impeller. The pump should be easy to take apart for cleaning and replacing worn-out parts.*

*Filter sponges. Wash these once a week.*

*Right: When installing the pump in the pond, turn off the electricity supply. Wrap some waterproof tape over the hose clip to prevent it from tearing the pond liner.*

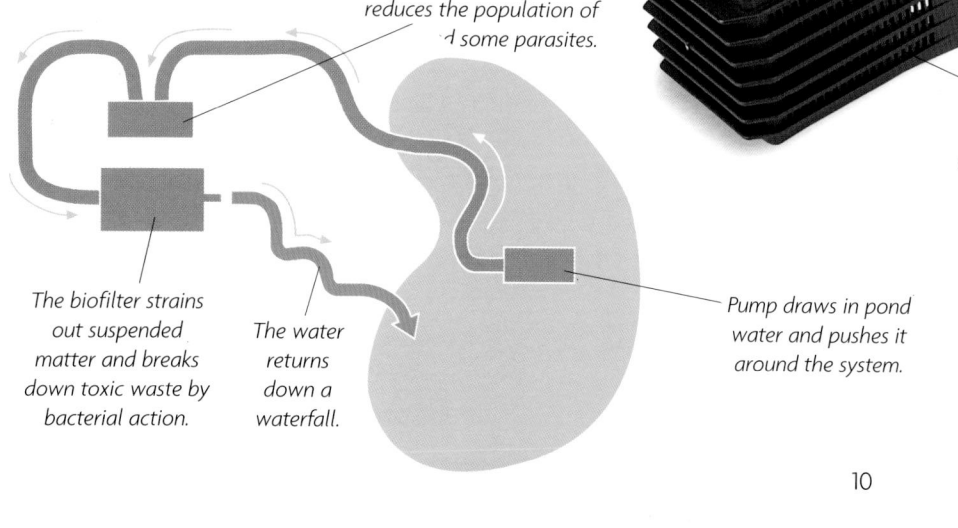

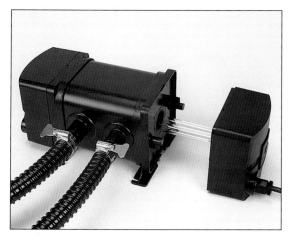

*Above:* Combined with a biological filter, an ultraviolet light (UV) sterilizer can replace hundreds of plants as a means of removing fish waste and breaking it down into harmless substances. The UV unit will also control green water and to a certain extent fish parasites.

## Maintaining the ultraviolet sterilizer

The UV sterilizer requires the least maintenance, but check it monthly to make sure the bulb is working. In most cases, bulbs need replacing once per year. (If you write the date on the bulb each time you change it, you will know when to replace it.) The best time to replace the bulb is in spring at the beginning of the new season. Remove the end section of the housing and pull out the bulb. Be sure to pull it out straight, otherwise you may break the quartz sleeve. Remove the sleeve and clean it at the same time. If it is dirty, the ultraviolet light cannot reach the water and will not work properly. If the wires are damp, spray them with a water repellent that is safe for electrical use. Never look at the ultraviolet bulb when it is working, as it will damage your eyes.

## Maintaining the filter

The biofilter is simply a collection point for fish waste. This is broken down in the filter, with the help of bacteria, into a nontoxic product (nitrate), which is used in turn by the plants as a source of fertilizer to aid their growth. The pump supplies the filter with water, oxygen, and dissolved waste, so it is important to keep the pump running 24 hours a day. In spring, it takes four to six weeks for the bacterial colony in the filter to reach full working capacity. Wash the sponges when they become clogged. How often this happens will depend on how dirty the pond is. Wash them in a bucket of pond water, as fresh tapwater might harm the beneficial bacteria. The biological media should not require any maintenance until the end of the season, when a quick wash in pond water will be enough.

This pipe directs the incoming water to the bottom of the filter box.

Different grades of foam trap floating particles in the water.

Cleaned water flows back to the pond from here.

Trays full of gravel hold down the foam layers and act as a second-stage biofilter.

These plastic pieces provide a large surface area for the growth of beneficial bacteria that "clean" the water.

As you get to know your fish, you will recognize their different characters and habits. It will be easy to see if a fish is ill or "off-color," as its behavior will be different from normal. It may stop eating or sulk somewhere away from the rest of the fish, or it may sit under the waterfall or fountain. If you observe any unusual behavior like this for a few days, it could signal a problem. Most problems stem from poor water quality, so this is the first thing to check out. If the water quality is good, the next step is to observe the fish carefully. Rapid breathing is a sign of parasite infestation and you will have to apply a general treatment to the whole pond. Always follow the instructions provided for using treatments and complete the course, even if the symptoms disappear.

Predators, such as herons and kingfishers, or even raccoons and cats, are a major problem for many pondkeepers. The only way to stop them from seizing your fish is to erect a barrier around the pond or to cover the pond with a net to prevent access to the water. Although not attractive, a net is the most effective deterrent. Plastic model herons are available but are of little use. If anything, they encourage young herons to fish in ponds. Planting the edge of the pond with tall plants will also deter herons, as they will not be able to reach into the water to fish, but you will still need a net during the winter when the plants have died down.

Like all animals, fish have a number of basic requirements, namely oxygen to sustain life, food to sustain growth and normal body functions, and a clean environment. If any one of these is lacking or inadequate and conditions are not corrected, fish will suffer and eventually die. The most common reason for pond problems is not checking the water quality, the second is buying poor-quality fish, and the third is overstocking the pond. This is not an exhaustive list, but these are certainly the most common reasons for pond failure.

### How many fish will my pond support?

*To estimate stocking levels use this simple rule: 2 cm (0.8 in) of fish length (not including the tail) per 50 liters (11 gallons) of water. For example, if your pond holds 1500 liters (330 gallons) of water it will support 1500 ÷ 50 x 2 = 60 cm (24 in) of fish. This means that you can keep, say, ten fish each measuring 6 cm (2.4 in) or five fish each measuring 12 cm (3.2 in), and so on. Never overstock your pond, as this can lead to poor growth and outbreaks of disease. It is much easier (and cheaper) to prevent disease than to cure it.*

**Below:** *The role that oxygenating plants play in the pond cannot be emphasized enough. As well as providing oxygen for the fish, they also remove minerals from the water, which discourages algae growth. To maintain the plants in top condition, remove the older parts and pot up the new growth. Oxygenators also provide food for fish, as they can eat the soft new growth, and the fronds afford shelter for young fish.*

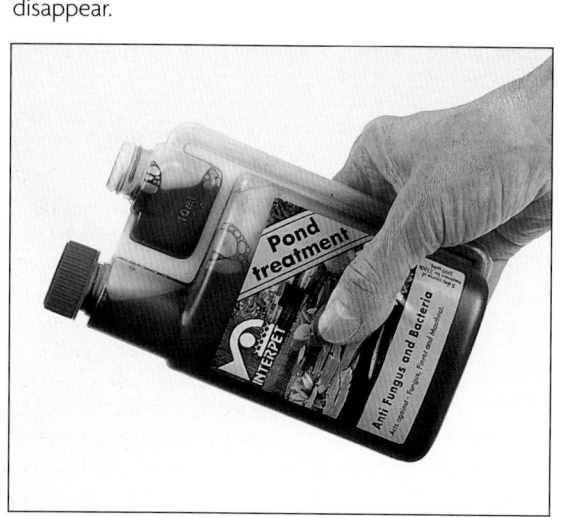

**Left:** *You can buy pond treatments for a wide range of conditions that might affect fish, such as parasites, and bacterial and fungal infections. These are supplied in easy-to-use containers that dispense a measured dose.*

Ceratophyllum demersum

## Testing the pond water

With the availability of modern test kits, checking water quality is a simple matter and you should do it on a regular basis, say once a month in the growing season and every two months in winter. Keep a record of the results so that you can compare them with previous tests. Once you have started to test the water, use the same test kits to be sure of obtaining comparable results. Different suppliers use different color schemes for their results.

The vital tests are for ammonia, nitrite, and pH levels (degree of acidity or alkalinity). Although there are other tests you can do, these are a good start and you can carry out more tests later if necessary. Tablet tests are better suited to beginners, as these tests are immune to "overdosing," unlike liquid tests. Tablets also have a longer shelf life – up to two years – whereas most liquids are only accurate for six months after opening. You can also buy paper strip tests that you simply dip into the pond or a sample of pond water.

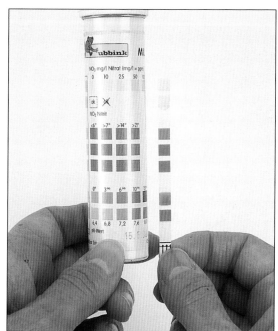

## Paper strip test

*These test strips are supplied in a container that has the reference color chart on the label. Take out a strip with dry hands and dip it into the water for one second and remove it. Wait for one minute for the colors to develop. Avoid shaking excess water off it, because this may mix up the results. Do not put used strips with new ones.*

*__Left:__ You can dip the test strip directly into the pond or scoop out a cup or jar of water to make the test.*

*__Above:__ Compare the strip colors with the chart. From the top these show: nitrate, nitrite, total hardness (three squares), carbonate hardness, and pH.*

## Testing for nitrite

*This is the second-stage product of the filter system and can be toxic in high levels over a long period. To reduce levels quickly, make a water change and stop feeding the fish. This stops them producing waste and gives the filter a chance to convert nitrite to nitrate.*

*__Right:__ Add one tablet to the water sample and shake. The color takes 10 minutes to develop fully.*

*__Below:__ Compare the color in the sample to the printed chart. This reading shows a high level of nitrite.*

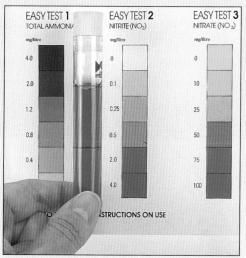

# Encouraging your fish to breed

Breeding fish can be difficult and prevailing conditions need to be suitable to encourage breeding. Some are more likely to breed if their survival is threatened, survival of the fittest being nature's way. However, to avoid threatening your fish, you can cheat. In late spring and early summer the fish will start to think about breeding and males often develop a spawning rash (white, sugar-grain-sized spots on the gills and along the front pectoral fin). When these appear, run the hose into the pond, as the addition of fresh water and the change in temperature can trigger spawning. It excites the males, who then chase the females around the pond until they spawn in the shallow water, normally on weed or marginal plant roots. If this does happen, you will have to rescue the eggs, otherwise the fish will eat some of them, but some will survive.

The best way to hatch the eggs is to place them into a fine mesh (2-3 mm/0.08-0.12 in) net in the pond, full of plants for the fry to hide in. The eggs should hatch in three to nine days, depending on the water temperature. The fry will not be quite ready for food, but within two days they should be ready for a first meal. A specially formulated liquid food for baby fish is available and is ideal for the first two or three weeks. Offer the fry small quantities four or five times a day. By the time they are a month old they will be able to tackle normal food, such as ground-up flake or pellets. At the end of the season, the young should be 4-6 cm (1.6-2.4 in) long and can be released into the pond with the adult fish.

Young goldfishes may not change color for up to two years, and some never do. Remove these uncolored fish (they are a dark bronze-brown) because they grow more quickly than the colored ones and when they breed they produce mainly uncolored youngsters, which you cannot see in the pond. As your pond becomes established, you will need to remove excess fish every two years or so, depending on how quickly your population increases.

*Below:* As the season moves on to early summer, the male fishes develop a spawning rash over the gill covers and front fins. This is the best way to distinguish between the sexes.

**Right:** You can feed fish fry with a liquid food supplied in a plastic dropper bottle that you simply squeeze to release the required amount of food. The liquid carries tiny food particles in suspension.

**Above:** The finely divided root systems of many floating plants, such as these on Pistia stratiotes (water lettuce), provide an excellent sanctuary for fry to escape predation – even by their parents.

**Left:** In a well-maintained garden pond, goldfish and many other types of pond fish will thrive and put on weight. During the summer months, well-fed fish often spawn and you may see tiny newcomers darting about.

# GOLDFISH • *Carassius auratus*

**Average size:** *20-23 cm (8-9 in) excluding fins, almost double including the fins on long-finned types*

The goldfish is the most common species in the garden pond world. It is a lively fish that is always on the lookout for something to eat. It is an excellent choice for the beginner as it is very hardy and can survive in most conditions. The basic bronze goldfish is the hardiest, while the more colorful varieties are mostly hybridized to obtain the color and finnage.

**Compatibility**

*All goldfish types mix well with other fish and should live for 20 years or more – less for the more hybridized varieties.*

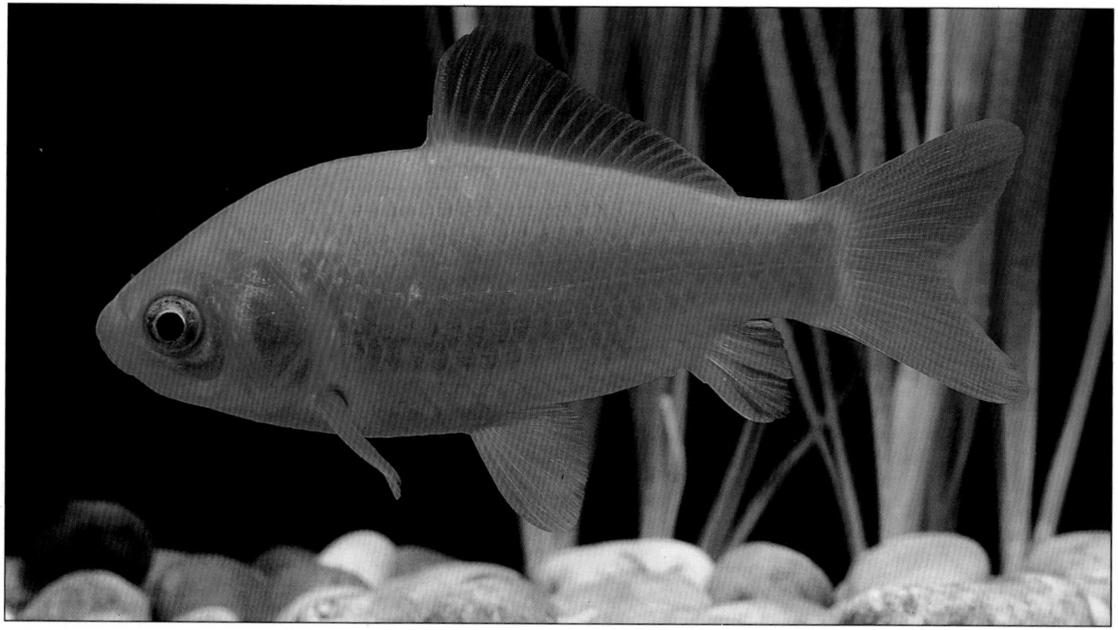

***Below:*** *Healthy goldfish keep their dorsal fins raised up and hold their other fins away from the body. Observe them regularly for any signs stress or disease.*

***Above:*** *Viewed from the side, the original goldfish shape is clear to see: a deep body and short, rounded fins. This is how the fish should look when you buy them.*

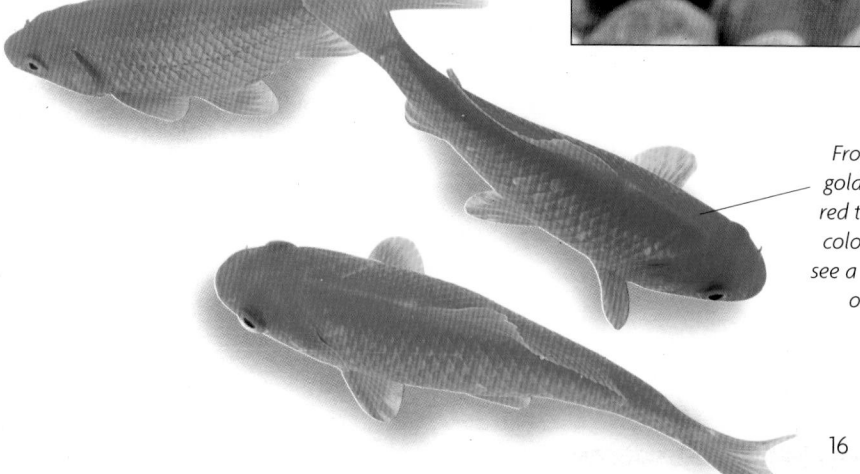

*From above, the goldfish is really a red to brick-orange color and you can see a few gold scales on the sides.*

## *Ideal conditions*

Food: Goldfish will eat almost anything, including live or dried foods. Brown bread is a favorite. Feed the fish little and often.

Minimum number in the pond: Goldfish are sociable fish, so keep them in small schools. Single fish will join schools of other species just to be in a group.

**Left:** This is the natural color of the goldfish and often the starting point for colored varieties. When the fry are born, their initial color can be influenced by temperature. In colder years, more fry are black/bronze colors and in warm years more are colored. Many fish are black and never develop into colored fish.

**Below:** Some fish start to change color in their first season, while others take up to two years to change. Depending on the weather, the fish can change color in anytime from a month to a year.

*Average size compared to this page*

The eventual size of a fish is governed by three factors: water quality, the number of fish in the pond, and the amount of food available. Inadequate conditions will have an adverse effect on the growth and health of the fish.

This young fish is in the process of changing color. At this stage, fish can look very pretty, but they do not stay this color for long.

# COMET-TAILED GOLDFISH

Comet-tailed goldfish, or comets, are much the same as normal goldfish. They usually have less body mass for the same size, but longer, flowing fins. However, they are not as hardy. With their large tail fins, comets are fast swimmers and very active in the pond.

*Below:* The comet's fins can be as long as the fish itself. When buying comets, check the fins for frayed edges (a sign of fin rot). Once settled in a pond, the fish's fins will regrow if they are damaged, but this can take a season or more.

**Sarasa comet**

*A good-quality sarasa comet is a very beautiful fish, with a crimson back, snow-white underside and long white fins. This fish is a "must" for the goldfish connoisseur. The best come from Japan, although other countries are now producing very good fish.*

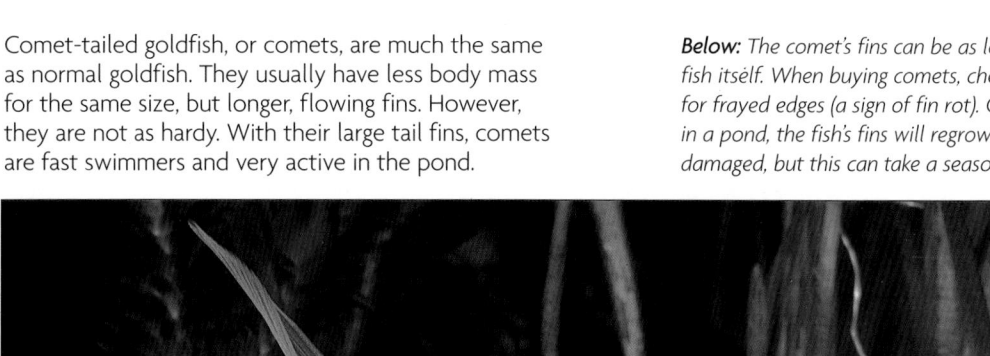

## ▶ *Ideal conditions*

Food: Offer the same diet as for the goldfish, namely a good mix of two foods, plus some frozen food such as bloodworm to add variety.
Minimum number in the pond: Keep comets in a group or with other types of goldfish.

# SHUBUNKIN

With its colorful body and long flowing fins, the shubunkin is probably the second most popular pond fish. It is most attractive and nearly as hardy as the goldfish. Two major varieties are normally sold in pet stores: the variety with smaller and more rounded fins comes from cooler climates, while the long-finned type is found in warmer climates. The fish from cooler climates are normally hardier as they can tolerate greater extremes of temperature.

*Below: The shape of the shubunkin belies its goldfish origins. If it is crossbred with goldfish, the resulting young revert to the normal bronze color. If you want to breed shubunkins, separate them from other types of goldfish.*

*Above: The long-finned variety is more commonly found in pet stores, as they are imported from warmer climates. As with comets, check the fins for damage. It can be difficult to see because of the fishes' coloring, so look carefully.*

## Ideal conditions

Food: Shubunkins will thrive on the same diet as described for other goldfish varieties. Provide a good mix of food with frozen foods as a supplement.
Minimum number in the pond: Keep shubunkins in a group.

# FANCY GOLDFISH

Fancy goldfish, originally bred in China, are suited to a quiet garden or patio pond, where their colors and finnage can be enjoyed without the distraction of other fish. In these conditions they have a chance to look and be at their best. It is not a good idea to mix them with more boisterous species that may hassle them and eat all the food. In warmer climates they can stay outside all winter, but in colder conditions they need the protection of an indoor pond or tank. Bring them in before the cold weather starts in autumn and do not put them out in the pond until the last frosts are over. In good conditions, fancy goldfish can live for 15 years or more. You can buy a special diet for fancy goldfish that is softer and easier for them to eat. Mix it with some frozen food.

Like all goldfish, they spawn easily, but the young grow slowly compared to "normal" fish, because hybridization over the years has weakened them. However, given enough food, a good number of the young fish will survive.

**Right:** *The fantail is the most popular fancy fish and available in many different colors and patterns. Another popular choice is the veiltail, which has longer flowing fins than other varieties.*

**Above:** *The white body of this fantail goldfish has a shiny, or mother-of-pearl, appearance and acts as a foil for the red-hooded area on the head. This marking – called a "red cap" or "tancho" – can occur on a wide range of goldfish varieties, but shows up best on white fishes.*

## Should I keep fancy goldfish in my pond?

*The simple answer to this question is: It depends on where you live. If you live in a warm climate without winter frost, you can keep a wide range of fancy goldfish in your pond. Even so, you will need to keep them in slowly moving water and offer them some protection from more boisterous fishes. However, some fancy goldfish are not suitable for ponds at all. These include the slightly bizarre telescope-eyed and bubble-eyed varieties. In these, the eyes are simply too vulnerable to damage to risk keeping the fish in any location other than an aquarium. Varieties such as the lionheads shown below can be kept in a pond in warm conditions or periods. The swollen growths on the top and side of the head form a brightly colored cap. This hinders the fishes' ability to feed, so check that they are getting enough to eat.*

The orfe, more often called the golden orfe, tuffy, or ruby red, is a selected form of the original color. The wild fish have a dark gray back and are silver-white underneath. Orfe are also available in blue and pink and can grow to 60 cm (24 in) or more. Although they look thin from above, they hide a deep body below.

To allow these fast-moving, active fish to reach their full potential, keep them in a reasonable school in a large pond. They like to swim upstream if they can, so waterfalls are a temptation they cannot resist if the flow is sufficient. Do not keep them in small ponds without a pump or air pump, as they prefer a high oxygen level.

## Ideal conditions

Food: These surface feeders will take most floating foods, snatching the food as they swim along. They also like to jump for flying insects. Minimum number in the pond: Five.

**Average size compared to this page**

### Compatibility

Orfe are a good addition to a quiet goldfish pond as they add movement, even in winter when the other fish are hibernating. However, it is best not to mix orfe with koi, as the orfe are somewhat nervous fish and prevent the koi from becoming tame. Furthermore, they cannot be treated with some koi remedies.

## Breeding

Orfe will breed when they reach 30 cm (12 in) or more. Breeding takes place in a fast-flowing area such as a waterfall. The fry are a large size when they are born, so survival rates are good.

# GOLDEN RUDD ● *Scardinius erythrophthalmus*

This truly active but nervous fish can unsettle other calmer fish, such as koi. Given plenty of swimming space, it is very successful in ponds and quickly multiplies. It is best confined to a sizable pond that can cope with a large number of fish. The fry are quite large when they are born and stand a good chance of survival. To keep the numbers under control, remove excess fish when necessary. It is most often seen in the golden form; as in the orfe, the natural fish is darker and less conspicuous.

*Average size compared to this page*

### ▶ *Ideal conditions*

Food: Golden rudd will eat any dried foods, but they prefer live foods, such as midge larvae and worms. Although a surface feeder by nature, the golden rudd can adapt to midwater and even bottom feeding.

Minimum number in the pond: Three.

While the golden rudd and the next three fishes are not particularly popular with pondkeepers in the United States, they have been included here because of their popularity elsewhere in the world.

*The golden rudd is often mistaken for the golden orfe, but it is darker in color. This is a very hardy fish that reproduces quickly.*

# TENCH ● *Tinca tinca*

The tench is one of the least-known and rarely seen pond fish. However, it is very useful in the pond, as it spends most of its time digging through the organic debris that builds up on the pond floor looking for food. This activity prevents the debris from decaying and poisoning the water, because it allows fresh water to reach all parts of the sediment.

For tench to thrive they need the security of a well-planted pond in which they can dash away and hide if necessary. They also require plenty of base cover to sift through. After a while they will become tame, especially if offered worms and other live foods. The fish can attain a length of 60 cm (24 in) and may live for up to 20 years.

*When buying new fish, check them over carefully. Look for sores on the underside; these are very common and can be a sign of disease.*

*Examine new fish to make sure they do not have sunken bellies. This indicates a lack of feeding.*

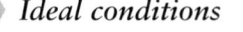

## Ideal conditions

Food: Tench will eat any dry food. They feed from the surface, but prefer sinking food and live food, such as worms or maggots. Other favorites are sweet corn and wholemeal bread. Minimum number in the pond: Three.

## Breeding

In the right conditions, tench start to breed once they reach 20 cm (8 in). The young have a poor survival rate, so they are unlikely to overpopulate the pond.

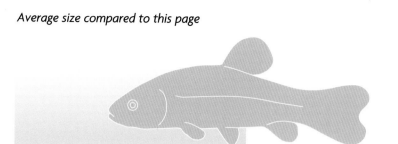

**Below:** *The tench is available in various colors, the most common being green, its natural color. The golden tench is one of the best-looking fish you can keep in a small pond and there is also a red-and-white variety from Japan. This color is rarely available, but a "must-have" if you find one.*

**Behavior**

The tench is often called the "doctor fish" because it can be seen rubbing itself against sick fish. In fact, it is actually transferring its own parasites to the sick fish, which are easier to parasitize than healthy fish.

**Compatibility**

Although they will coexist with any fish in a planted pond, do not keep tench with too many other fish as they tend to be shy feeders and the more boisterous fish get to the food first. They cannot compete with koi in an unplanted pond.

*Black coloration on golden tench is common and not a sign of problems or disease.*

Since it only grows to 10 cm (4 in), the minnow is an excellent fish for patio ponds and small raised ponds. It is active by nature, but unlike other active fish, it does not unsettle other fish as it is so small. In fact, it mixes very well with smaller fish. Nor does it upset the plants in a planted pond, as it is a midwater feeder looking for insects. It does best in schools of five or more because it feels more secure in a group. It breeds well in ponds, providing there are not too many larger fish to eat the fry. It makes a good substitute for golden orfe in small ponds. It does not live as long as some coldwater fish, but you can expect it to survive for up to five years.

*Average size compared to this page*

**Right:** *A small school of minnows will often feed from your hand in a madcap fashion, especially if you offer them bloodworm. Small children find them highly entertaining!*

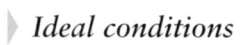

**Above:** *The minnow can be difficult to find in shops. It is often sold as a dace, so check before you buy so that you know exactly what you are getting.*

▸ *Ideal conditions*

Food: Preferably live or frozen foods, but the fish will accept dry foods, such as flake and micropellets. Minimum number in the pond: School of five or more.

The roach is another example of a coarse fish that has found its way into the ornamental pond. It is not quite as nervous as the rudd, but like that fish it can breed at an alarming rate if not checked. In spring, the head of the male is covered in white spots, known as a breeding rash. This is the best time to buy the fish as you can select all males or all females to prevent them from breeding. Roach are better suited to a wildlife pond, as they are not particularly colorful. Their average length is 20 cm (8 in) but they can grow to twice that size.

**Average size compared to this page**

*The roach and the rudd are often confused, but you can tell them apart by the position of the pelvic fins. In the roach, the fins start in line with the dorsal fin, but in the rudd they start before the dorsal fin.*

### ▶ *Ideal conditions*

Food: Roach will eat any floating food and live foods.
Minimum number in the pond: Small school of five or more.

The bitterling is not often offered for sale, but it is a good fish for a small patio or goldfish pond, as it does not grow too large, reaching just 10 cm (4 in). It will mix with any fish and, being an active species, it provides movement and interest. The male is considerably more colorful than the female, often showing metallic red, blue-purple, and green coloration, especially during the breeding season. Provide a well-planted pond with some clear swimming areas.

*Males are more conspicuous in color than females. In spring, the colors increase in intensity and males will spar with each other to gain the right to breed with available females.*

**Average size compared to this page**

### Ideal conditions

Food: The bitterling's preferred diet is small insects, but it will take flake food and minipellets.
Number in the pond: Three or four pairs.

*To encourage the fish to breed, add some freshwater mussels to the pond.*

### Breeding

The fish breed by laying their eggs in freshwater mussels for protection. When they are about four weeks old, the fry are released from the mussel. Since the breeding method is unusual, they will not overrun the pond.

In some parts of the world, the sterlet is a relative newcomer to the pond fish market. When buying it, take care not to confuse it with the Siberian sturgeon *(Acipenser bareri).* The latter is also often sold as a "sterlet" but grows to 2 m (6 ft) or more and is clearly not suitable for a garden pond. Even *A. ruthenus* is a sizeable fish that can reach 16 kg (35 lb) in weight. To distinguish between the two, remember that the sterlet has white edges to its fins and nose and a darker gray-colored skin. The sturgeon has a brown-tinted skin and no white edges to its fins, but it quite often has a white nose.

**Average size compared to this page**

**Compatibility**

*The sterlet will mix with any fish, but has been known to eat small tench if they do not move out of its way as it searches the bottom for food.*

### ▶ *Ideal conditions*

Water: Sterlets are active from dusk till dawn and during colder weather. They do not like high temperatures because these lower the oxygen content of the water. Provide flowing water and plenty of swimming space.

Food: These bottom-feeders need good-quality sinking food as a staple diet; surface-feeding is very difficult for them to cope with. For best results, offer them high-protein sinking food available from trout farms or specialist fish retailers. The sterlet will also eat chopped worms, fish, and snails.

Minimum number in the pond: One in a small pond, two or more in a larger pond.

# KOI or COMMON CARP • *Cyprinus carpio*

***Average size:*** *Males and females 60 cm (24 in)*

Koi are the most varied colored fish that you can keep, occurring in all colors from golden-yellow to sky blue. Each color variety has a Japanese name, as it was the Japanese who first bred colored carp hundreds of years ago. Koi are lively, potentially large, permanently hungry bottom-feeding fish that quickly learn to feed from the surface. They spend most of their time looking for food. Koi like plants – to eat! If bought as young fish and allowed to grow up with plants, they sometimes leave them alone, but this is rare. Most plants will be dug up and spread around the pond. Koi can live up to 50 years or more, but generally survive for 10 to 20 years in a garden pond. They are truly good pond fish, providing you have the space and filter systems to grow them to maturity.

***Average size compared to this page***

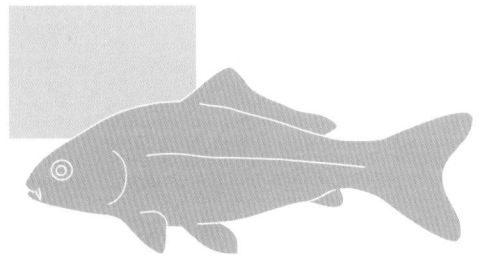

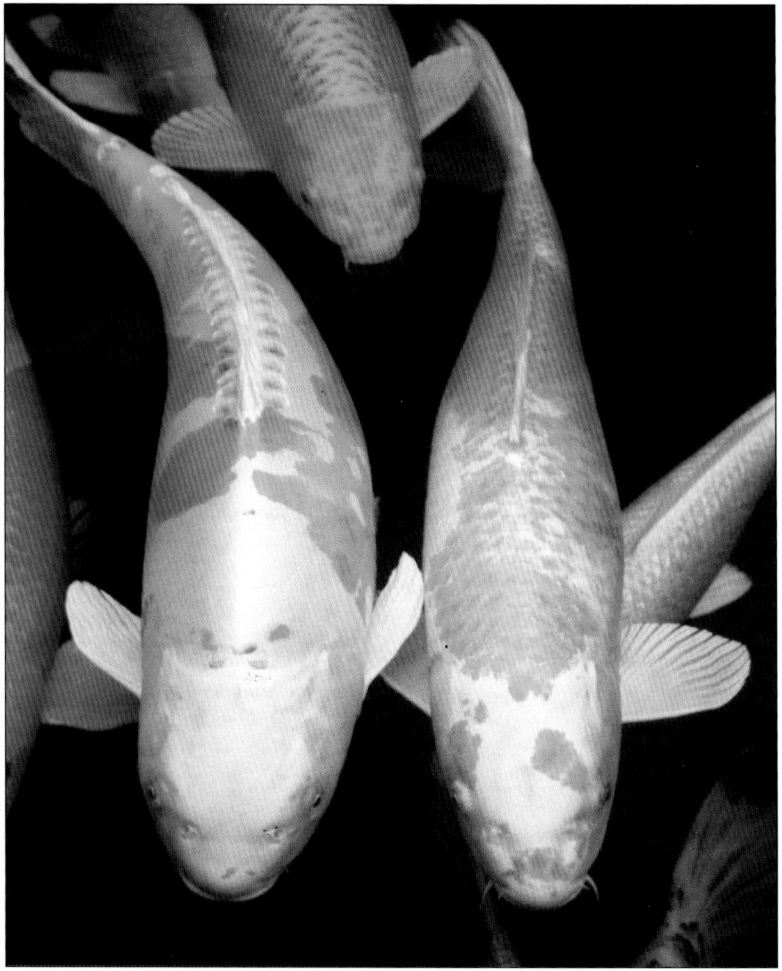

## *Ideal conditions*

Water: One of the most tolerant fish species. They will survive in almost any water conditions or temperatures. Providing the pond is deep enough, koi will live through most winters.

Minimum pond size: Koi are ideal for a large, deep pond of 10,000 liters (2,200 gallons) or more. They can be kept in smaller ponds, but soon outgrow them.

***Left:*** *Koi are social fish that prefer to be with other carp. They often swim around the pond following the largest fish. They feel safe in numbers and can be coaxed into hand-feeding more easily when there are several of them in the pond.*

## The common carp

The common carp is the natural form of the koi and has the same requirements, but it lives longer and will grow larger. Like koi, it will dig up all the plants in the pond and eat them. However, it makes a good addition to a pond without plants, because being a calm, good-natured fish, it will encourage other fish to hand-feed.

*Below:* This is the fully scaled common carp. Although it is not very visible in the pond due to its dark coloration, once it has grown to a good size you will spot it instantly at feeding time as it vacuums up all the food it can find!

 ## Breeding

Once mature, koi breed easily if rather boisterously, but will eat their eggs as fast as they lay them, so remove the eggs as soon as possible.

*Right:* Koi feed as quickly as they can and then spend time chewing the food as they swim around the pond. Once they are finished, they come back for more. Make sure all the fish get something to eat by spreading the food around.

# Feeding koi

Koi have huge appetites and must have good-quality food. If any minerals or vitamins are missing in their diet, it can cause slow growth or poor health. Fish are fed by the weight of food not by volume, so feed sticks in greater quantities than pellets.

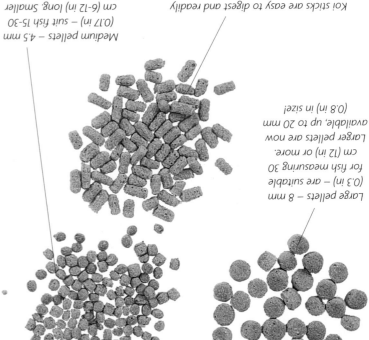

Koi sticks are easy to digest and readily accepted by most fish. They are a good way of encouraging smaller or new koi to feed, but provide them in sufficient quantities.

Large pellets – 8 mm (0.3 in) – are suitable for fish measuring 30 cm (12 in) or more. Larger pellets are now available, up to 20 mm (0.8 in) in size!

Medium pellets – 4.5 mm (0.17 in) – suit fish 15-30 cm (6-12 in) long. Smaller fish can eat at this size, but it is better to feed the right size pellet for the fish.

Special foods are available to improve color and encourage growth. Many include spirulina, an algae that enhances the skin and natural colors of the fish. Feeding rates vary, so read the instructions.

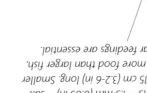

Small pellets – 1.5 mm (0.05 in) – suit small fish 8-15 cm (3.2-6 in) long. Smaller fish require more food than larger fish, so regular feedings are essential.

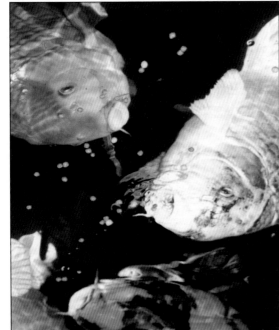

**Below:** Koi will accept a wide range of foods and in large quantities. A basic diet of koi pellets, supplemented with brown bread, boiled potatoes, sweet corn, and lettuce is a good starting point.

**Above:** *Given the chance, koi will sit midwater in the sun and sunbathe. This can lead to sunburn, so it is a good idea to provide some shade in the form of overhanging plants so that they can find some shelter if they need it.*

**Left:** *Advances in filtration equipment mean that you can be sure of having clear water in the pond without adding any plants. However, to make up for this lack of vegetation in the koi's diet, you should offer them a supplement of lettuce and other greenfoods.*

## Compatibility

*Koi are social fish and do not worry other fish, although very small fish may be eaten accidentally when the koi are surface feeding, as they cannot see in front of their mouths.*

33

**Right:** An early stage in the first round of judging at a koi show. Here, the judges vote on the merits of an individual koi "bowled" into a plastic floating basket for closer inspection. The judges work as teams, each team being responsible for judging different-sized groups throughout the day.

## Showing koi

One of the main reasons that people keep koi, other than for pleasure, is to compete in shows. It is a very competitive area that often involves acquiring expensive fish in order to compete at national levels. However, local shows are a lot of fun and a good way of judging for yourself what it is you want to achieve from the hobby. Shows are also a good place to meet fellow enthusiasts.

When selecting fish for showing, look for clear definitions between the colors. They should be strong and even over the body. Often the head color is a different shade, say darker. Bear in mind that as young fish grow they are very unlikely to retain the color or pattern they had when you bought them. If you want to show fish, it is normal to buy larger specimens to begin with, say 25 cm (10 in) or more in length. That way there is a greater chance that the colors will remain close to the desired pattern. As you start to look for better-quality fish, the cost of buying them increases and the reason is quite simple. Koi produce many young, but as breeders apply the exacting standards demanded of top-quality fish, fewer and fewer make the grade. The time and expense involved in the breeding process are inevitably passed on to the customer.

# Popular color varieties

**Below:** Ogon is the number-one seller for most shops as it is a bright yellow color and, being less hybridized, is hardier than most colored koi. It often grows to become the largest koi in the pond and one of the friendliest. Look for a clear yellow head without orange or murky brown areas. Generally speaking, large fins are an indication that the fish will grow to a good size.

**Above:** The red and white koi, called the kohaku, is the most popular patterned koi and the starting point for most collectors. A good specimen should have snow-white skin and brick-red patterning down the body. The red should not run into the tail or down over the eyes. Kohakus with a deep cherry-red color usually develop black coloration as they age, so avoid buying them.

**Below:** At one time, the showa was a black fish with some red and white coloration, but in the "modern" showa there is a more even split between the colors and a lot less black than in the original variety. The black should cover part of the head and all the fins should be colored or striped, as here. The red should be as red as possible.

The elegant flowers of Iris laevigata provide stunning color along the edge of a pond.

# Introduction

# *Choosing pond plants*

It is true that plants are the decoration for the pond, but this is to understate their value and the important role they play in maintaining water quality and balance. Although many ponds are equipped with ultraviolet sterilizers, biofilters, and other electronic equipment, people are beginning to realize that plants still have an important part to play and are introducing them to ponds in greater numbers. It is not generally realized that plants can remove toxic products from both the pond water and the tapwater that is used to fill the pond. However, the advantages of adding plants to the pond soon become apparent. Fish become more active, their health improves, as does their growth rate and often their color. Best of all, the green film of hair algae stops growing and dies away. It is difficult to overplant the pond, but once the water "disappears," then lack of space is clearly a problem. It is relatively easy to overcome this by growing some plants in a stream and/or waterfall and not all in the pond itself. This also makes managing the plants easier, as they are more accessible. The more you cut back the plants, the quicker they grow, and the more they take up waste material from the water, the cleaner the water becomes.

Read up as much as you can before buying any plants, as some can grow very large and be unsuitable for smaller ponds and gardens. In fact, some water plants are so vigorous that they are classed as weeds in certain countries; check local regulations. On the following pages, you will find a selection of plants from oxygenators to water lilies. Achieve the right "mix" and you can create a beautiful, well-ordered pond.

# Using plants in a pond

**Right:** Water lily leaves and oxygenating plants provide a dense cover in this established garden pond. A good balance of aquatic plants will keep the water clear of algae and offer shelter for fish and other wildlife living in and around the pond.

When planting your pond, the most important factor to take into account is the effect that plants will have on the water quality. In most clear, natural ponds there is a generous surface coverage of aquatic plant leaves – often floating plants, such as frogbit or duckweed. Unless you intend to have a fish-only pond (say, a pond for koi, which eat or disturb any plants), there should be a large proportion of submerged and floating plants to start the pond.

The key to maintaining a good balance in the garden pond is to maintain a high ratio of plants to fish. The waste produced by the fish will encourage green water caused by microscopic algae. These will reproduce in vast quantities if the ornamental plants cannot absorb the mineral byproducts quickly enough and offer some shade at the water surface. The first plants to introduce are the oxygenators and surface-leaved plants, such as water lilies. As both types grow from the base of the pond to the surface, some people recommend placing a layer of soil across the liner for them to root into. Unfortunately, this rarely works; one or two of the more vigorous plants usually take over and smother the more desirable ones, thus producing a large, uncontrollable green jungle in the pond.

# The role of plants in and around the pond

### Oxygenators

These are the first plants to go into a new pool and offer shade for the fish and food if needed. As they grow quickly they control the nitrate levels and check the algae growth. Plant them in baskets to make removing excess growth easy.

Elodea crispa is one of the best oxygenators and quickest to establish.

### Floating plants

Floating plants provide cover in a new pond until the deep water plants are established. They offer instant cover and shade for fish and reduce the amount of sunlight reaching the water. Their roots are also a haven for young fish.

Azolla (fairy moss) covers a large area quickly and is easy to remove as it spreads.

### Deep water aquatics and water lilies

These offer the main shade in the established pond and grow in the deeper water where other plants would not survive. The large leaves, especially those of water lilies, offer a safe area for fish to rest and hide away from predators above.

Nymphaea "Texas Dawn" is one of the best water lilies available for a larger pond.

### Marginal plants

Marginals offer more than just color and form, although this is their main role. Their other "duties" include absorbing minerals from the water and creating shade and a convenient route for the wildlife to get into and out of the pond.

With its distinctive banding, zebra grass offers height and color to any pond edge.

### Bog garden plants

Bog plants thrive in the transition zone between the pond and dry land. In natural ponds the plants in this moist environment hold back the banks with their roots. Around garden ponds they can provide color and interest at every season.

Lobelia "Vedrariensis" bears these purplish-violet flowers in summer.

# Planting a pond basket

When planting a selection of aquatic plants in the same basket, do not mix strong-growing varieties with weak ones, otherwise the weak ones will be overrun. Be sure to use good-quality aquatic soil and not peat-based potting mixes. These float in the pond and release a brown color into the water.

*A selection of pond plants.*

*Planting baskets are available in a range of shapes and sizes.*

*Planting baskets with fine holes mean you do not have to use burlap or foam liners to retain the soil.*

*Good-quality, lime-free aquatic soil.*

*Slow-release fertilizer tablets.*

*Lime-free gravel. Add a layer of gravel to the planted basket.*

*1* Place a layer of aquatic soil in the bottom of the basket. When you stand a plant on it, the top of the soil of the plant should be about 2.5 cm (1 in) below the lip of the basket.

*2* Remove the first plant from its pot. Take off any weeds or dead roots and put it in position. Arrange the remaining plants in the same way, leaving room for them to grow and spread.

*3* Carefully fill in around the plants with more aquatic soil and firm it down well. Choose plants with a range of height, color, and shape to add interest to the planting in the pond.

4 Place one long-life fertilizer pellet next to each plant. These should last the whole season and encourage the plants to grow and flower.

5 Top off the basket with fine, rounded, lime-free gravel. Fish can damage themselves on sharp gravel and it could puncture the liner.

Carex riparia "Bowles' Golden"

Myosotis scorpioides

Iris laevigata

6 Water the planted basket thoroughly before placing it in the pond. This will help to remove any loose soil and prevent it from being released into the pond.

### Placing a water lily in the pond

Before choosing a water lily, be sure to take into account the size of the pond, the water depth and the amount of sun it receives. Choose the variety carefully, because planting an unsuitable one may cause problems.

1 If the leaves do not float on the surface, use a clean, upturned plastic pot to raise up the lily to the correct height. A black pot will be less visible in the water.

2 Thoroughly water the lily before you place it in the pond. Add a generous layer of gravel so that the fish cannot disturb the plant as it becomes established.

**Left:** Water lilies, such as this Nymphaea "Amabilis," are excellent pond plants. The leaves provide shade for the pond and a welcome refuge for fish. The flowers are available in a wide range of colors, from white through yellow, pink, and red.

# Oxygenating plants

The oxygenators are the first plants to introduce to the pond. They are generally sold in bunches tied with lead at one end. This helps them to sink to the bottom, but the best way to establish them is to remove the lead and plant them in baskets with aquatic soil and gravel in the same way as other marginals. Like most plants, they require some soil to root into and in poor-quality water they can extract nutrients from the soil not found in the water. Only plant one type of oxygenator in each basket, as the dominant one will displace the others. If you have a small pond and wish to establish it quickly, start with *Elodea canadensis* and remove it when the more desirable oxygenators have become established in the water.

**Above:** *Plant oxygenators, such as* Elodea crispa, *in baskets topped with gravel. Lower the basket to the bottom of the pond and let the plants grow to the surface. To trim back excess growth, lift out the basket.*

## ▼ Callitriche verna (C. palustris)
Starwort

*Star-shaped rosettes of leaves float on the surface and a thin root descends to the bottom of the pond. Very small green flowers appear in late summer. It is a very good oxygenator but not suitable for natural ponds, as it will take over. However, grown in a container of heavy clay soil in cold, deep water, it will flourish.*

## ◀ Ceratophyllum demersum
Hornwort

*This oxygenator can be quite difficult to grow as it likes still, nutrient-rich water. However, once established, it will grow quickly and you will need to remove excess growth. This is easy to do, as the plant forms no roots but just floats around the pond. Many fish like to spawn on it, so if you intend to breed fish, try some. Pinch out the tips in winter and leave them in the pond. In spring, they will grow into new plants. Remove excess plant material before the cold weather sets in.*

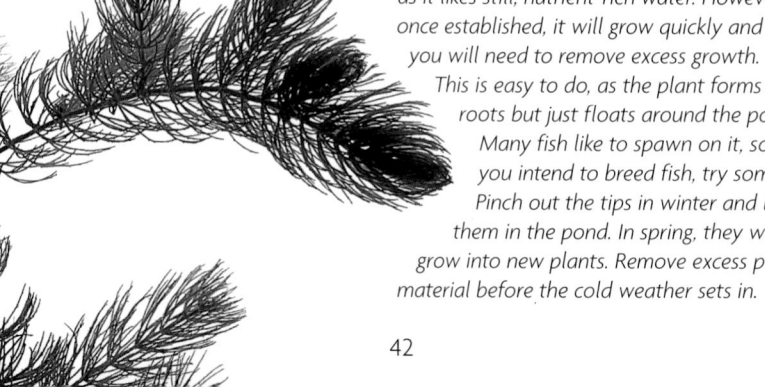

## Elodea canadensis
### Canadian pondweed

Without doubt, this is one of the best – if not the best – oxygenators you can plant. In good conditions it can grow at an alarming rate, but as long as it is contained in pots it is easy to control. Do not plant it in clay-bottomed ponds, as it takes over. Fish enjoy eating the younger shoots. Trim the stems back to about half every month.
This will stimulate new growth. Only trim one basket at a time so that the others can "work" while the new growth appears.

## Ranunculus aquatilis
### Water crowfoot

Water crowfoot is an odd plant, as it has two types of leaves: a feathery submerged one and a clover-leaf type on the surface. In early summer, small white flowers with yellow centers are held just above the water surface. The only drawback is that the plant dies back after flowering, but this is a small price to pay for a flowering oxygenator. It is not invasive and a good plant for shallower ponds.

## Elodea crispa
### Fish weed

Elodea crispa is another very quick grower and should be contained. As the plant matures it becomes untidy, so for best results, propagate new plants every two years or so. Pinch out the top 4 cm (1.6 in) of growth and pot the cuttings up in shallow water in spring. As they root and start to grow, remove the older plants and replace them with the new ones.

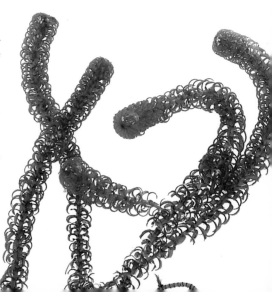

### ► *Eleocharis acicularis*
Hairgrass

Often sold as a tropical plant, hairgrass grows well at the margins of ponds, and in its correct place on the bottom, it is a reasonable oxygenator that produces an underwater lawn. It is very easy to propagate; simply divide it and pot it up in spring. Fish love spawning on it in the shallows and it provides an excellent nursery for the fry.

### ▲ *Myriophyllum aquaticum*
Parrot's feather

*Myriophyllum* needs lime-free water to flourish. The feathery bronze green flowers are held high above the water. M. aquaticum (parrot's feather) is the best-known variety and by far the quickest grower. It heads straight for the surface and grows across it. It can become a pest, but is good for providing cover in new ponds. Remove it when the more desirable plants have established themselves.

The lower leaves die off, leaving only the foliage above water. Cut back the dead plant material every year before the first frosts so that it does not pollute the water.

The feathery bronze green foliage spreads in loose mats beneath the surface and in summer small

## Hottonia palustris
### Water violet

Although not a particularly good oxygenator, the very attractive water violet more than makes up for this shortcoming with its bright green feathery foliage and spikes of violet-colored flowers in early summer. It is not an easy plant to establish, but worth a try. Fish tend to dig it up, so give it some protection until it attains a reasonable size.

Once established, hottonia forms a bright green carpet just above the surface. From here, it sends up the small violet-colored flowers.

## Utricularia vulgaris
### Greater bladderwort

This is a good oxygenator and one of the few insectivorous plants you can buy for the pond. It requires good conditions to do well, namely warm, acidic water and a good supply of small invertebrates, which it captures in the submerged bladders. In late summer, small golden-yellow flowers are held above water. This is a very interesting plant and if you can grow it, so much the better, as it is becoming rarer each year.

### Other oxygenators

Potomageton crispus
This excellent oxygenator, which looks more like seaweed than a freshwater plant, does best in moving water. As it never goes wild and takes over the pond, it is a good choice for people with limited time to spend on maintenance. In shady ponds the soft foliage is green, but in sunny areas it turns bronze.

Fontinalis antipyretica
The feathery fronds of this plant – willow moss – grow on rocks. It prefers moving water in streams or deeper waterfalls, so it is not suitable for still ponds.

# Floating plants

Floating plants provide instant cover for new ponds until the deep water plants have grown enough to create shade. Most floating plants grow quickly and need regular thinning out so that they do not end up shading the submerged plants they are supposed to be helping. A few of them are not hardy in all but the warmest climates and need protection. Alternatively, replace them each year.

## Hydrocharis morsus-ranae
Frogbit

*Although late to start and early to fade, frogbit is a very good floating plant. Small, round, fleshy leaves begin to spread out in early summer to be followed later by white flowers. By early autumn, the frogbit drops resting buds to the bottom ready for the following year. This is a short season by any standard, but the small lily-like foliage and flowers are a welcome addition, especially to small ponds.*

## Azolla caroliniana
Fairy moss

*This aquatic fern is not hardy in cold regions but survives by sending spores to the bottom to overwinter. In very sunny or cold weather, the leaves turn a bright brick red. It can take over smaller ponds, but is easy to control by removing excess growth from the surface with a net. It is a useful plant for new ponds as it provides quick surface cover, but it can become a pest.*

## Stratiotes aloides
Water soldier

*Water soldier looks like a pineapple top and has similarly serrated edges. It prefers cold water. In summer, it floats to the surface and by midsummer may send up white flowers. Once finished, it sinks to the bottom, where it acts as an oxygenator, and roots for the winter. It reproduces by sending out runners that break away when mature.*

*Use* Trapa natans *as a pond bedding plant, rather than as a perennial, and replace it every year.*

## ▲ Trapa natans
### Water chestnut

Water chestnut is not hardy in cold regions. It rarely flowers, but if it does, and the white flower is pollinated, it forms a small black fruit that sinks to the bottom. This waits until spring before germinating to form a new plant. In most years, it is easier to over-winter some plants in a warm greenhouse and then place them in the pond after the spring frosts.

## ▶ Pistia stratiotes
### Water lettuce

The green corrugated leaves of this tropical floating pond plant rise 15-20 cm (6-8 in) out of the water. It does best in calmer waters away from waterfalls, otherwise the leaves tend to be pushed underwater by the ripples. If you want to breed fish, the long feathery roots provide excellent cover for developing fry. It grows quickly in a warm summer, but transfer it to a frost-free greenhouse for the winter in cold areas.

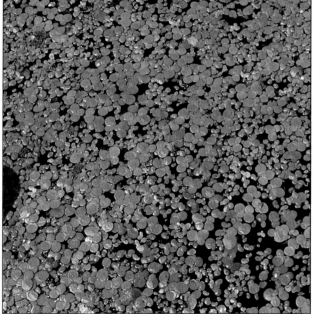

## ◀ Lemna
### Duckweed

Duckweed grows at an alarming rate and, once introduced into a pond, is almost impossible to get rid of. However, it provides virtually instant cover, conceals fry from potential predators in a rearing pond and controls hair algae (as it excludes the light). Remove excess growth every week.

Although water lilies are the typical surface-leaved plants, there are other deep water aquatics that you can use. They have the same requirements as water lilies and, when planted with lilies, they extend the flowering season and add interest to the pond. They are often better suited to difficult situations and will grow where water lilies would struggle, say, in shade or moving water. Planted in rich soil, they soon make their presence felt in any pond and, being hardy, will survive cold winters.

▶ *Aponogeton distachyos*
  Water hawthorn

*The water hawthorn sends up sweet-smelling flowers as early as late winter and continues to do so until late autumn, with a small break in midsummer. This makes it a good plant to mix with water lilies, as they flower in midsummer. The large oval leaves provide cover for the fish, and shade in the summer helps keep the pond cool. Very easy to grow and one of the few pond plants with scented flowers.*

**Right:** *A well-established water hawthorn can often flower for the entire season, as it produces more and more rhizomes in the planting basket. Remove fading flowers to encourage new blooms to develop.*

▲ *Nymphoides peltata*
Water fringe

The fringe water lily is ideal for situations where water lilies will not grow, such as ponds with moving water or fountains. Given the chance, it will quickly cover the surface; remove excess foliage regularly. Small yellow flowers appear in early summer. Remove dead flowers to stop them setting seed and to encourage the plant to produce more flowers into the autumn.

◀ *Orontium aquaticum*
Golden club

The golden club is often overlooked, as people are reluctant to wait until it reaches maturity. However, it is very hardy and deserves a place in the water garden. Once mature, large, spear-shaped leaves surrounding long white stems with golden spikes at the tip appear in early spring.

# Marginal plants

Often dismissed as having little or no influence on the pond environment and only good for their aesthetic value, marginals are underestimated. As well as offering height and shade around the pond, they also allow access to and from the pond for wildlife. But the major advantage of these plants is their ability to absorb excess minerals in the pond water more quickly than the slower-growing submerged plants. On the aesthetic side, they extend the flowering period in and around the pond, and the range of shapes and colors increases the overall visual effect that you can create.

Most marginals are similar to herbaceous plants and can be treated as such. Pot them in good-quality, lime-free aquatic soil and add a long-term fertilizer pellet with low nitrogen to encourage flowering. Cover the soil with rounded gravel to stop the fish from excavating the soil. Plants will need dividing every four to five years or sooner if they outgrow their welcome in the pond. Always keep the young shoots for repotting. Some of the more invasive plants can be grown in solid plastic pots to stop excessive root growth from moving into neighboring pots.

**Right:** A well-planted pond enhances any garden or yard, and the reward for spending a little time each week maintaining it is a fine show of plants, such as this *Pontederia cordata*.

▶ *Acorus calamus* "Variegatus"
Variegated sweet flag

The variegated sweet rush has green-and-cream striped foliage and very small aromatic flowers. Only the coldest weather causes the foliage of this semi-evergreen plant to die down. It is a good plant for the darker corner, since it will keep its color even in shade.

▼ *Butomus umbellatus*
Flowering rush

The flowering rush has rich green leaves that are triangular in section and can reach 90 cm (36 in) in height. The clusters of pink flowers in late summer are even taller, although it can be some time before the plant is mature enough to flower. It does best in full sun and shallow water up to 4 cm (1.6 in) over the soil surface.

▲ *Alisma plantago-aquatica*
Water plantain

The water plantain produces a rosette of large, spoon-shaped leaves with well-defined veins running along their length. Throughout the summer, a pyramid of white, sometimes pink-tinted flowers appears on a stem rising 60 cm (24 in) from the center of the leaves. It is best to remove the spent flowers before they set seed, as the plant can spread quite quickly across any damp ground.

► Caltha palustris
Marsh marigold

The marsh marigold, or kingcup, is a must for any pond, as it is the first pond plant to flower in spring and one of the most prolific growers. Buttercup-yellow flowers on stems up to 45 cm (18 in) long are held above a mound of glossy green leaves. Cut it down in midsummer to encourage new growth; old leaves can become mildewed and untidy.

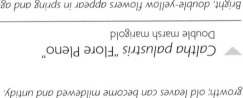

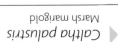

▲ Caltha palustris "Flore Pleno"
Double marsh marigold

Bright, double-yellow flowers appear in spring and again in autumn. It is one of the better-known varieties and a good choice for the smaller pond, as it is compact and does not have the excessive growth of C. palustris.

▲ Calla palustris
Bog arum

White arum flowers in summer are followed by red berries in autumn. The glossy heart-shaped leaves are produced on rambling stems. This is a very good plant for the bog garden and pond edge as it offers a thick matlike entrance ramp that frogs and toads can use to reach the pond.

## Carex riparia "Bowles' Golden"
### Golden sedge

The golden sedge is not really a water plant, but adapts readily to the pond environment and also does well in a bog situation. It is an excellent plant for the darker corner of the pond, as it retains its color, even in the shade. In late summer, dark brown flower spikes appear above a mound of golden leaves.

## Cyperus longus
### Sweet galingale

The sweet galingale is a truly graceful plant, with three-sided leafy stems that bear long, reddish-brown flower spikes. It is best reserved for larger ponds, as it grows to 120 cm (48 in). However, if planted in a large basket it can make a striking impact in the smaller pond.

## Cyperus alternifolius

The umbrella palm is an architectural plant, with thin green stems up to 60 cm (24 in) tall that support star-shaped formations of almost lime-green foliage and rather insignificant brown flower spikelets. It seeds itself in or out of the pond and spreads quickly. It can be tender in very cold climates, so drop it into deeper water for the winter to protect it from frost.

## Houttuynia cordata "Chameleon"

The leaves of this plant are most unusual, as they may be red, green, yellow, or white in no particular order or combination. The same plant will vary in color at each season and from year to year. In addition to the riot of leaf color, it also produces small white flowers in late summer. It will grow in any soil from wet to dry and can be invasive if not controlled.

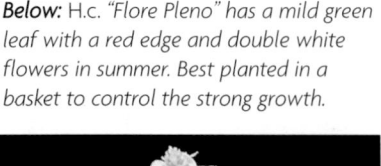

**Below:** H.c. "Flore Pleno" has a mild green leaf with a red edge and double white flowers in summer. Best planted in a basket to control the strong growth.

## Glyceria maxima var. variegata

This striking plant has beautiful green-and-creamy white-striped leaves that are flushed pink when young. Like most grasses it can be invasive, but its growth can be controlled by planting it in a basket. It is best not planted in a clay-bottomed pond.

## Cotula coronopifolia
Brass buttons

The fine, bright green foliage makes a lush mat in spring. Once the small, bright yellow, buttonlike flowers are over, the foliage may deteriorate, but after a quick trim new growth will soon appear and the plant often flowers again.

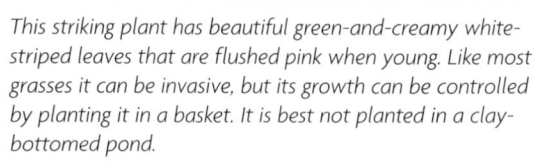

▲ *Iris laevigata* "Variegata"
   Asiatic water iris

This is one of the best variegated plants for the pond, as the cream and green colors are well separated. The blue flowers are short-lived in late summer, but when mature, the group will offer a good flowering period. It is less vigorous than other irises and a good choice for the smaller pond. Needs lime-free soil for best results.

▶ *Iris pseudacorus*
   Yellow flag

The flag iris is a very strong grower, and its free-flowering nature and large yellow flowers make this the most impressive iris. The dark green foliage up to 100 cm (39 in) tall creates an imposing plant in any but the smallest pond. Confine it to a basket for best results and to control the root system. A variegated version is also available.

▼ *Iris versicolor*
   American blue flag

This iris is ideal for the smaller pond. Its restrained habit and narrower leaves make for a more compact plant, although it still grows 60 cm (24 in) tall. The violet-blue flowers are held on 45 cm (18 in)-high stalks. Does best in shallow water 2-4 cm (0.8-1.6 in) over the roots.

## Lobelia cardinalis ▲

This is really a herbaceous plant, but does well in water away from its arch enemy, the slug, which seems to eat the entire plant overnight. It is a very striking plant, because even before it flowers, the dark red stems and foliage are immediately visible around any pond. The flowers are also a strong crimson-red and appear on long thin stems up to 1 m (39 in) tall. As they open a few at a time, the flowers will seem to last for months.

## Lythrum salicaria ▲

The purple loosestrife is an excellent choice for the pond. Thin green leaves are held below tall spikes of pink flowers that flower from midsummer until early autumn. There are a number of varieties, ranging in height from 45 to 150 cm (18-60 in). L. s. "Robert" has clear pink flowers held about 75-90 cm (30-36 in) above ground. It is ideal for most ponds.

## Juncus effusus "Spiralis" ▶

Unlike its relatives, which have straight spikes, this strange plant produces spirals of green cylindrical leaves growing in all directions. It is a novelty, but a good conversation point, as everyone will comment on it.

## Lysichiton camtschatcensis
White skunk cabbage

If you have the space this is a must. Once mature, the immense foliage is just part of the effect, forming the backdrop to the enormous white arum spathe and green central flower spike. Not as large as the yellow L. americanum. Needs a very large container of rich soil to do well. Feed in the growing season.

## Lysimachia nummularia
Creeping Jenny

Bright green leaves covered in butter-yellow flowers make a cheerful addition to the pondside or rockery and soften the edge of a waterfall. This highly versatile plant will grow anywhere and over anything and is a firm favorite with pondkeepers.

## Mentha aquatica
Water mint

The water mint is a very fast-growing plant that will soon take over the pond if not controlled. But it is worth keeping for its small, aromatic, oval leaves and purple "balls" of flowers that attract a host of insects during the summer. Cut it back after flowering and it soon repays you with more flowers and foliage.

## Mimulus
### Monkey musk

The monkey musk is one of the earliest plants to flower, and if deadheaded can flower all summer and into the autumn. If you allow it to set seed, it soon appears everywhere. The best hardy variety is M. guttatus. Its large yellow flowers with dark red spots over mid-green leaves brighten even the darkest corner. Other varieties are available, but not all are winter hardy in temperate regions.

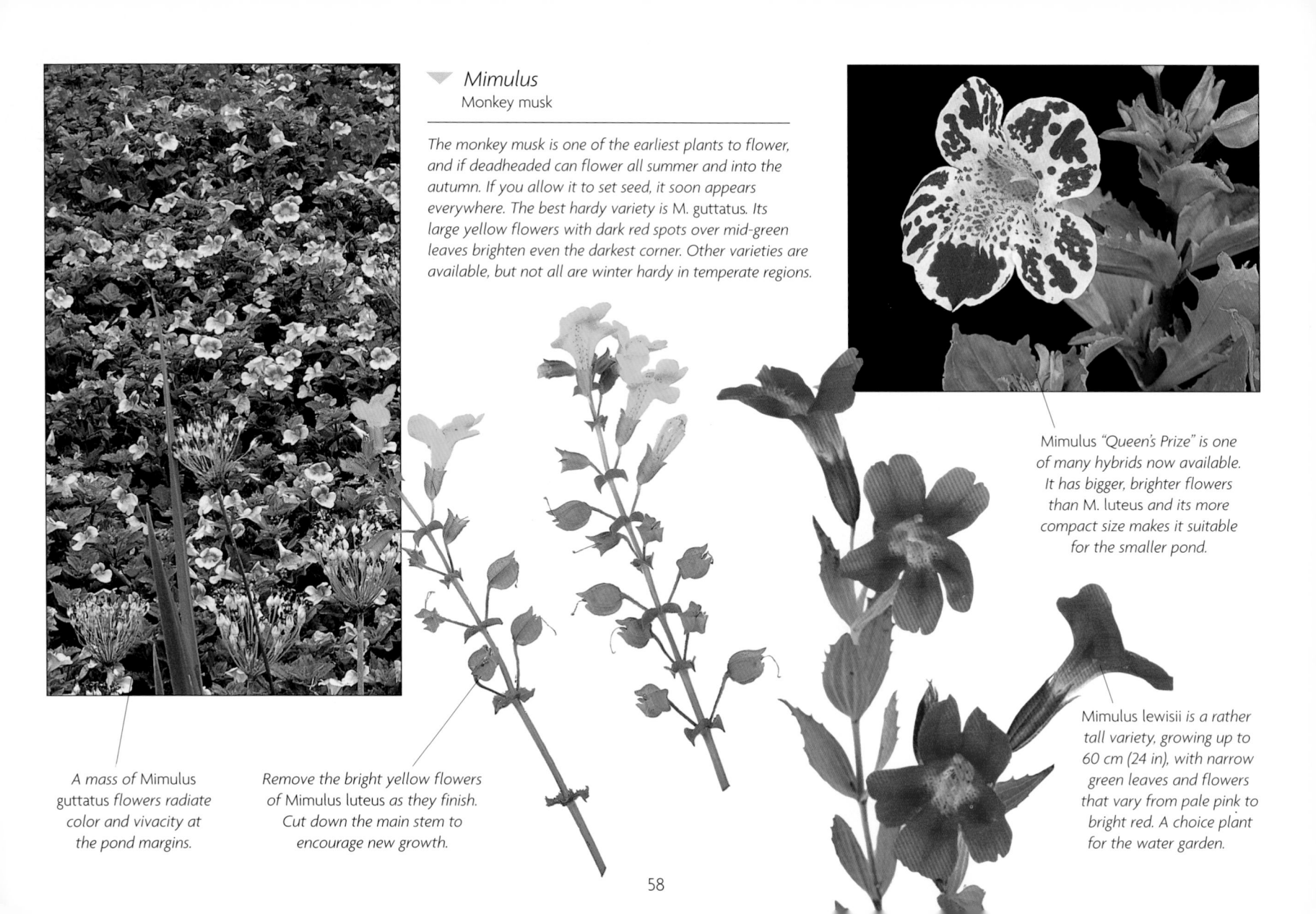

Mimulus "Queen's Prize" is one of many hybrids now available. It has bigger, brighter flowers than M. luteus and its more compact size makes it suitable for the smaller pond.

A mass of Mimulus guttatus flowers radiate color and vivacity at the pond margins.

Remove the bright yellow flowers of Mimulus luteus as they finish. Cut down the main stem to encourage new growth.

Mimulus lewisii is a rather tall variety, growing up to 60 cm (24 in), with narrow green leaves and flowers that vary from pale pink to bright red. A choice plant for the water garden.

## Nasturtium officinale
### Watercress

This is a very useful plant in the pond. Used correctly, it can reduce algae growth almost to nothing. As it grows so quickly it uses up any excess nutrients in the water and starves the algae. Crop it regularly to encourage new growth. Plant it in baskets without soil so that the plant has to extract the nutrients from the water rather than from the soil. Cut off flowers as they go over, otherwise it will set seed and die down. Ideal to grow in a waterfall.

## Myosotis scorpioides
### Water forget-me-not

With its small, pale blue flowers held close to the water surface, the water forget-me-not is familiar to most people. It has a rather straggly growth habit, but if trimmed forms a good mat in the margins. Look out for M. s. "Mermaid," an improved hybrid that is even more free-flowering and has larger leaves. (The pink form here is M. s. "Pinkie.")

## Oenanthe javanica "Flamingo"

Grown for its fernlike foliage of gray-green leaves tipped with pink and cream, this plant can be invasive if it escapes to the garden. However, in the pond it creates a colorful display all year. Pinch out flowers as they finish to keep the foliage in good condition. Not ideal for the wildlife pond, where it may escape, but good for smaller ponds.

*The blue flower spike rises through the heart-shaped leaf in late summer. A white variety is available.*

### ◀ Pontederia cordata
Pickerel weed

Despite its common name, this is one of the finest plants for the pond margin. It is attractive in every respect, from the glossy green, heart-shaped leaves to the spikes of clear blue flowers that appear in late summer and into the autumn. In a large basket in full sun it will grow to 60 cm (24 in) or more in height. Repot it regularly for best results.

### ▲ Phalaris arundinacea

This graceful grass has an invasive habit, so grow it in a container to stop it spreading across the pond and garden. Cut it down each spring to encourage the younger growth, which has a more vivid color.

### ▲ Ranunculus lingua
Water buttercup

The water buttercup is only suitable for the larger pond, as it can grow to a height of 120 cm (48 in). The long spear-shaped leaves and large yellow flowers from spring until autumn will grace any pond. It is unusual, as it grows in all but the coldest winters, and so provides some greenery when most other plants have died down. For the smaller pond choose R. flammula, which has small yellow flowers held in bunches 45 cm (18 in) high.

## Sagittaria sagittifolia "Flore Pleno"
### Arrowhead

As its common name suggests, the glossy mid-green leaves of this plant are shaped like an arrow. They offset the triangular stems of double white flowers grouped in threes at intervals along it. To see it at its best and to encourage flowering, plant it in a rich, acid soil in shallow water. S. sagittifolia has single flowers with a purple eye.

## Rumex sanguineus
### Bloody dock

This foliage plant has blood red veins running though the glossy green leaves. As the common name suggests, it is related to dock, but is considerably better looking and makes an interesting addition to the margin of any pond. A recommended plant.

The double white flowers in summer enhance the value of this plant in the water garden.

## Saururus cernuus
### Lizard tail

The lizard tail plant has a long "tail" of tiny, white, fragrant flowers and olive-green, heart-shaped leaves that are brighter green on the underside. A good plant for the pond, as it is one of the few marginals to have autumn color; at that time of year the leaves turn a bright crimson.

▲ Typha latifolia "Variegata"

Together with Typha minima (shown above), this is the only reedmace to consider, as the more common ones are far too invasive and grow too large for most gardens. This variety also grows tall – up to 1.2-1.5 m (48-60 in) – but not as quickly and is still only suitable for the larger pond. The beautiful pale yellow variegation is very attractive and makes a vivid display.

▲ Schoenoplectus lacustris "Albescens"

This elegant, tall, green-and-white, vertically striped rush produces a dense growth in a few seasons. Repot or feed it every year to avoid short, thin growth. Cut it back in winter to about 15 cm (6 in) and retrim it in spring to remove older growth and encourage new shoots.

▲ Schoenoplectus lacustris "Zebrinus"

Another rush to look for is S. l. "Zebrinus." This striking plant has horizontally banded stems in green and white. It creates a tight clump as it grows and makes an interesting addition to the pond.

Collect the seedheads before they set seed to avoid an uncontrolled spread of new plants.

### ▶ *Zantedeschia aethiopica*
Arum lily

The arum lily is one of the easiest water plants to grow. In good soil and shallow water, it produces its instantly recognizable, dark green, horn-shaped leaves. And in late spring/early summer, the white flowers (actually spathes) with a bright yellow spadix, appear in the center. In winter, submerged under 25 cm (10 in) of water, it will be protected in all but the coldest climates. However, remember to raise it up to shallower water in the spring. Alternatively, place it in a frost-free greenhouse for the winter.

### ▼ *Veronica beccabunga*

This low creeper is ideal for the pond edge, as it grows just 10 cm (4 in) tall. Small blue flowers that resemble forget-me-nots appear above the rich, dark green, oval leaves. It will grow anywhere, even in fast-flowing water, so it is perfect for the waterfall area or alongside streams.

### ▶ *Typha minima*

Although often called a bulrush, this plant is in fact a reedmace and ideal for the small pond. Its thin 2-4 mm (0.08-0.16 in)-round, dark green leaves contrast well with other marginals. It will grow in the most exposed areas without any shelter, but contain it in a pot, as its hardiness means it will outgrow weaker plants. In autumn, small, almost round, brown seedheads form and these soon seed if left alone. If picked before they mature, they make good dried flowers.

## Bog garden plants

Poolside planting complements the pond and helps to blend it into the garden by slowly graduating from bog plants to garden plants. Because the soil warms up more quickly than the water, bog plants are the first to grow and flower, thus adding interest to the pond before the water plants are ready to grow. It is easy to add a bog area to the pond, even if the main pond is complete. Bury a simple liner with drainage holes 38-45 cm (15-18 in) underground. Make it a saucer shape to hold water under the plants and then plant into the soil above. The liner will hold extra water when the weather is dry.

*Right: Primulas are some of the first plants to flower in spring, adding color at an otherwise dull time of year around the pond. These are* P. beesiana. *Leave the heads to set seed, since they do not flower twice in a year.*

### ◀ Gunnera manicata

The enormous "rhubarb" leaves, measuring up to 3 m (10 ft) across, are a bit of an overkill for most people, so try G. tinctoria, a dwarf version, instead. The large green leaves and red stems, both covered with large spikes, make this one of the most impressive pond plants and an essential addition if you have space. Cover it in winter to protect it from frost and provide rich soil for the best growth.

### ▶ Primula vialii
Orchid primula

The orchid primula is probably the most photographed flower by a pond – and rightly so. The red-budded flower spike, held high over green leaves, opens to display mauve flowers.
In a group they are an impressive sight, especially as they bloom early in the season.

## ▲ Iris sibirica "Ewen"

This is a very hardy iris, but it does require some time to settle in. The purple-blue flowers with white specked throats are produced in late spring to early summer. The iris prefers neutral to acidic soil that covers the roots to 2 cm (0.8 in). Mulch compost around the crown in spring and again after flowering.

## ◀ Astilbe x arendsii
False goat's beard

Astilbes are popular bog plants that grow in moist, neutral to acidic soil in semishade. From early summer to early autumn, feathery flowerheads in colors from purple to red and pure white are held above green pinnate foliage. A hardy plant, ideal for windy positions.

## ▼ Trollius europaeus
Globe flower

The globe flower is a hardy perennial, with foliage divided into smaller leaves. The erect stems bear between one and three clear yellow flowers in early summer. The globe flower will grow in any soil, from damp to very wet, and in sun or shade. With the advent of hybrids, new colors are becoming available, but these are not as hardy.

Because they are large plants, lilies are often thought to be very hardy, but this is not so. For best results, always buy lilies established in aquatic baskets, with their roots emerging from the sides. Choose small lilies in 3-4 liter (3-4 qt) pots, medium ones in 5-10 liter (5-10 qt) pots and large ones in 10 liter (10 qt) pots or larger. The rhizome should be hard to the touch and the leaves strong and ridged. If they are withered and soft, it means that the plant has been allowed to dry out.

Transport the lily in a plastic bag with the top tied to prevent dehydration. Once home, add a fertilizer tablet and extra gravel over the soil. Place the lily in the pond, with the leaves 4-5 cm (1.6-2 in) below the surface. If it is too deep, the plant will waste energy growing to the surface and be weak. As the leaves reach the surface, lower the plant until it is at the correct depth. Never exceed the maximum recommended depth.

Do not feed lilies every year, as this will encourage leaves rather than flowers. Replace the soil every two to three years with good-quality lime-free aquatic soil. Most lilies will require splitting up every four to five years, depending on the variety.

### The sacred lotus (Nelumbo nucifera)

*The sacred lotus (Nelumbo nucifera) is one of the most spectacular water plants and ideally suited to warm climates. However, in temperate climates it can be grown outdoors for the summer and then overwintered in a conservatory pool or greenhouse. It grows 1.5-1.8 m (5-6 ft) tall, with leaves up to 90 cm (36 in) across and its large size can dwarf many ponds.*

### Nymphaea "Gonnère"

*A truly stunning plant, with large, white, double "chrysanthemum" flowers over round green leaves. It will grow in water up to 1 m (39 in) deep, and is ideal for medium-sized ponds.*

### Nymphaea odorata minor

*This is the smallest of the white lilies and ideal for tubs and small patio ponds. Small green leaves surround pure white, star-shaped, freely produced flowers. These are sweetly scented, adding to their attraction for a raised pond.*

## ▲ *Nymphaea* "Marliacea Albida"

*A free-flowering scented lily with clear white blooms held above the water surface. The golden-yellow stamens are quite conspicuous, while the large leaves are dark green with a purple underside. Needs a medium to large pond as the plant can spread out to 2 m (6 ft).*

## ▶ *Nymphaea alba*

*The common water lily in Europe, with large green leaves and pure white flowers. It is very hardy and ideal for deep ponds. Although often sold for small ponds, it is not suited to them at all, as it grows too large too quickly.*

# Pink water lilies

### ▶ Nymphaea "Marliacea Carnea"

A strong-growing lily for large ponds. Its large leaves and free-flowering nature make it a popular choice for growers and private gardens alike. The flowers are pinkish-white and improve with the age of the plant.

### ▼ Nymphaea "Firecrest"

If space allows, this superb lily with bright coral-pink flowers and yellow stamens is the one to have. In 60 cm (24 in) of water, it will cover an area of water 1.2 m (4 ft) across. Start it off in shallower water for the first year to encourage the rhizome to grow.

**Variable water lilies**

Most lilies in this group have yellow to orange flowers that gradually change to deep orange or red. They are suitable for the warmer positions in the pond and thrive in shallow water. They are prolific bloomers and their changing colors add to their value in the pond.

**Below:** N. "Sioux" is one of the variable colored lilies available. The new flowers open to a bright apricot and deepen to red-orange with age. The large blooms are produced in good numbers and the leaves are spotted brown or purple against a green background. A good specimen for the medium to large pond.

### Nymphaea "Laydekeri Lilacea"

An excellent choice for the shallower pond, as it prefers a water depth of 15-30 cm (6-12 in). The flowers open with a soft rose color and mature to a rose-crimson, with bright yellow stamens in the center. A very popular plant for smaller ponds and patio tubs.

The elegant blooms are set off against large mid-green leaves that can reach 25 cm (10 in) across.

### Nymphaea "Madame Wilfon Gonnère"

One of the best pink water lilies. The double flowers are a soft pink with a white flush toward the edge, and plain green leaves complement the flowers in summer. It grows well in deeper water and can spread to 1.5 m (5 ft) in diameter.

### ▼ Nymphaea "James Brydon"

A medium lily with peony-shaped, double, pinkish-crimson flowers. It is an old and very well known variety. Although rightly popular, it can be too vigorous at times, but planting it in poorer soil can encourage flowering and may slow down leaf production.

### ▶ Nymphaea "Pygmaea Rubra"

A very ornate, free-flowering lily for the small pond or tub. It is easy to grow and quite hardy for a small lily. Because it is slow to propagate, you may have problems finding it; many pet stores offer other plants under this name. Buy a flowering plant or go to a specialist.

## Nymphaea "Froebelii"

This reliable, free-flowering lily has the most perfect blood-red flowers with orange stamens, and what is more the flowers are scented, too. A good choice for the smaller pond.

## Nymphaea "Escarboucle"

A beautiful large-growing lily with very deep carmine-red flowers and large round green leaves. It is easy to grow and the perfect lily for a large pond.

## Nymphaea "Laydekeri Purpurata"

Crimson-red flowers with bright orange stamens are held above maroon-flecked leaves that are purple underneath. It is quite a prolific water lily that will flower for most of the season. Ideally suited to a shallow pond (say with water 15-25 cm/6-10 in deep) in a sunny position.

# Yellow water lilies

## Pests of water lilies

The few pests that appear each year are simple to control without the use of chemicals, which would harm any fish in the pond. Water lily aphids are a common pest. The small black aphids appear on the leaves. Wash them off with a water jet as often as possible and the fish will eat them. Spray nearby fruit trees, as the aphids winter in them. The water lily beetle is another problem. The black grubs and dark brown beetles eat the water lily leaves and flowers. Wash them off and remove all plant material from the pond, as it dies off in autumn. Finally, look out for the china mark moth. The caterpillars cut oval pieces out of the leaves and stick to the underside of the leaf. Pick off the larvae by hand and dispose of them.

▶ *Nymphaea* "Pygmaea Helvola"

*This is the smallest yellow lily and best suited to very shallow water, with a maximum depth of 25 cm (10 in). Although small, the red and green leaves are colorful and the canary yellow flowers are numerous, even on young plants. It is ideal for a patio pond or tub. Keep it in full sun for best results.*

## Nymphaea "Odorata Sulphurea"

Scented yellow flowers are held above the water surface, in a similar manner to the tropical lilies. The attractive leaves have almost chocolate-colored markings on top with red on the underside. Although small, it flowers well and does best in a smaller and shallower pond.

## Nymphaea "Texas Dawn"

If you can find it, this is the best yellow water lily. The flowers are flushed peach at the base and offset by large red, speckled leaves. It is only suitable for larger ponds, but if you have the space, it is a must.

## Nymphaea "Marliacea Chromatella"

This is the most popular yellow water lily, although not the best one and often unreliable. It grows large and has good flowers set against foliage variegated in green-and-red.

## Tropical water lilies

Tropical water lilies are often overlooked by pond-keepers in colder climates, but provided you can winter them in a greenhouse or conservatory maintained at 15°C (60°F) you can safely grow them. Do not transfer tropical water lilies outdoors until the water temperature exceeds 21°C (70°F). They will thrive in a sunny, sheltered spot. Bring them indoors in early autumn and keep them in shallow water in a light position. Tropical water lilies are ideal for shallow, warm water outside in summer — say in a patio pond — and they offer colors not available in the hardy lilies, such as blues and violets. Many have serrated edges to their leaves and hold their flowers high above the water surface. Always read the label when buying tropical lilies, as many flower at night.

◀ *Nymphaea* "Haarstick"

*The very dark cerise-pink to red flower opens at night and closes by morning. The only way to admire it is by flashlight, but it is well worth it. Otherwise, the most you can hope to see are the green leaves with serrated edges.*

▲ *Nymphaea* "Mrs. G. Hitchcock"

*Another night-blooming water lily, but this one has a pale pink flower with a yellowing center held high above the water surface.*

▶ *Nymphaea* "Albert Sibert"

*This day-flowering lily has pink blooms that deepen in color toward the center. The flowers are held above the leaves. It does best in shallow water ponds in full sun.*

▼ *Nymphaea* "Tina"

*"Tina" is a beautiful lily with dark blue flowers at the tips, often with a purplish tinge at the base. A small lily ideal for a patio pond. It blooms during the day.*

▲ *Nymphaea* "Mrs. Pring"

*The pure white flowers of this tropical water lily open during the day. It is quite a prolific grower.*

# INDEX

Page numbers in **bold** indicate major entries; *italics* refer to captions and annotations; plain type indicates other text entries.

**CREDITS**

The publishers would like to thank the following photographers for providing images, credited here by page number and position: B(Bottom), T(Top), C(Center), BL(Bottom Left), etc.

The Beaver Collection © Sue Westlake-Guy: 44(R), 45(T), 47(BC,TR), 51(C), 52(BL), 54(BR), 55(TL), 59(BR), 61(C), 62(C), 63(BC), 66(B), 67(TL), 68(BL), 69(L,R), 70(R), 73(TR,BR)
Eric Crichton: Credits page, 36, 38, 55(R), 57(L), 59(BL), 60(TL), 62(L), 63(R), 64(BL)
Frank Lane Picture Library: 26(B) © Gerard Lacz, 27 © W. Meinderts/Foto Natura, 31(L) © W. Meinderts/Foto Natura
John Glover: 48(R), 49(L), 53(L), 55(C)
S & O Mathews: 43(TR), 49(R), 58(TL), 60(BC), 63(T), 65(TC)
Clive Nichols: 52(BR), 65(R,L)
Nishikigoi International Nigel Caddock: 31(R), 33(TR), 35(L,C,R)
Photomax (Max Gibbs): 6, 15(TR), 16(T), 17(T,B), 18(L), 19(B), 22, 23, 24, 25, 26(T), 29
Geoffrey Rogers © Interpet Publishing: 8, 9, 10, 11, 12(L), 13, 15(TL), 32(L), 39(BL, BCL), 40, 41, 42(TC), 43 (BL,BR), 46(L), 47(TL), 51(R), 52(TC), 54(TC,TR), 57(C,R), 58(TR), 59(C), 60(R), 61(R)
Geoffrey Rogers: 16(L), 56 (C)
Fred Rosenzweig: 19(T), 20, 21(BR)
Mike Sandford: 21(L), 28
Neil Sutherland © Geoffrey Rogers: Title page, 12(R), 39(CB,BCR,BR), 41(BR), 42(BL,BR), 44(TL), 45(BL), 46(TC,BR), 48(L), 50, 51(BL), 53(C,R), 54(L), 56(L,R), 58(C), 61(L), 62(R), 63(L), 64(R), 66(TR), 67(TR), 68(T,BR), 70(L), 71(TC,TR,BL), 72, 73(TL), 74(L,R), 75(L,TR,BR)
W A Tomey: 14, 18(R)
David Twigg: 30, 31(R), 33(L)

The artwork illustrations have been prepared by Stuart Watkinson and are © Interpet Publishing.

Thanks are due to Anthony Archer-Wills; Blagdon Garden Products Ltd., Bridgewater, Somerset; The Dorset Water Lily Company, Halstock, Dorset; Heaver Tropics, Ash, Kent; Hozelock Cyprio, Haddenham, Buckinghamshire; Neales Aquatic Nurseries, West Kingsdown, Kent; Phoenix 2000, Pinxton, Nottinghamshire; Washington Garden Centre, West Sussex for their help in the preparation of this book.

The information and recommendations in this book are given without any guarantees on the part of the author and publisher, who disclaim any liability with the use of this material.